Tracy Nelson Maurer

A Crabtree Seedlings Book

Crabtree Publishing
crabtreebooks.com

TABLE OF CONTENTS

WINGS OF WONDER

Bats are the only **mammals** that can truly fly. Mammals have fur or hair on their bodies. Mammal babies feed on their mother's milk.

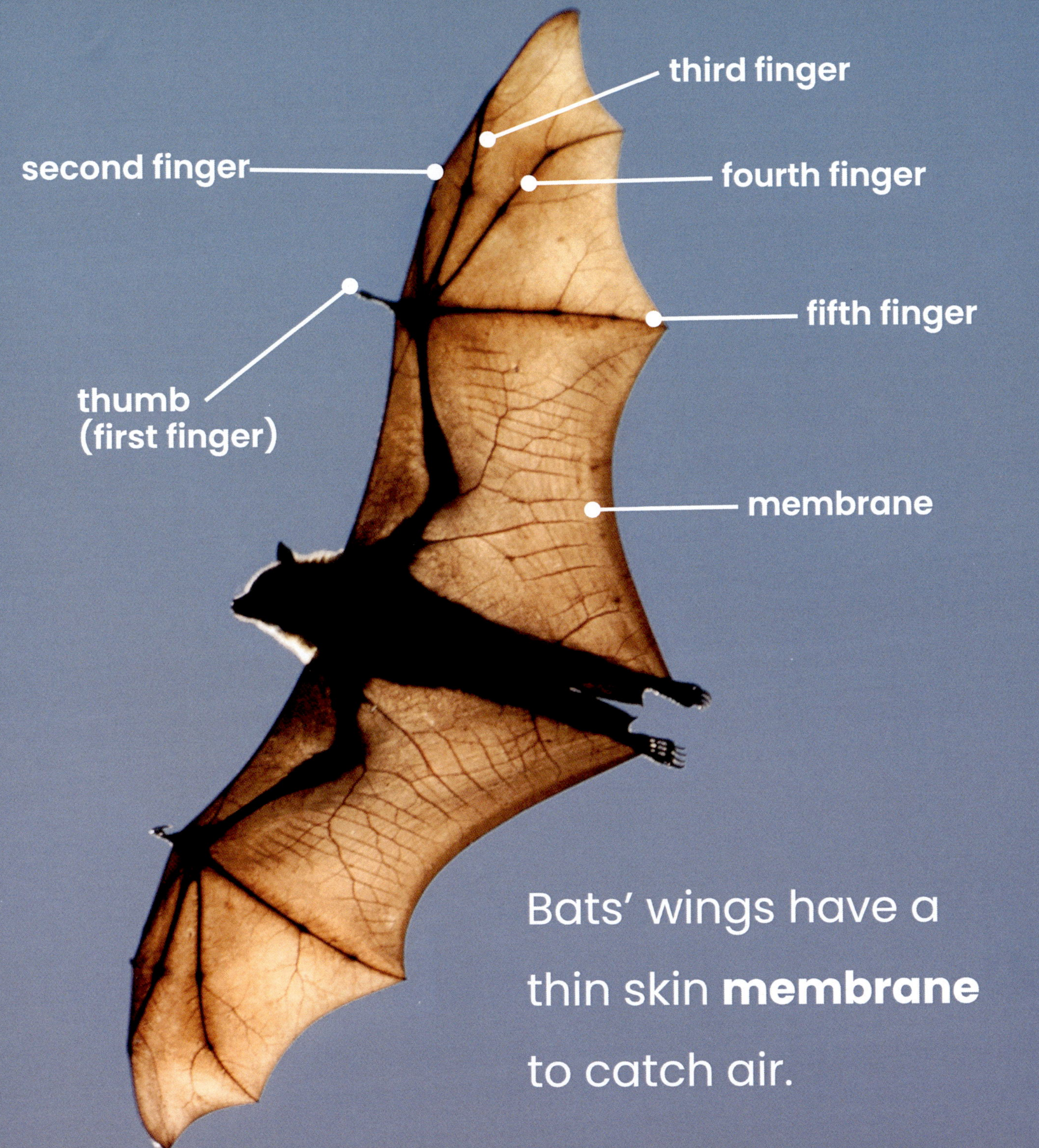

Bats' wings have a thin skin **membrane** to catch air.

Most bats cannot take off from the ground. To fly, they drop off a ledge, tree, or other high spot.

Bats move their wings the way humans wave their fingers. Bats also have a thumb for gripping.

NIGHT DINERS

Most bats are **nocturnal**. They fly at night to find food and they sleep during the day. Most bats eat insects, such as mosquitoes, beetles, and moths, or small animals such as frogs.

Persian trident bats eat insects.

Some bats eat fruit.

They live in the **tropics**.

The greater bulldog bat has pouches in its cheeks to carry fish when it flies.

Wrinkle-faced bats suck on over-ripe bananas.

Bats may look scary, but they are helpful! Fruit bats spread seeds and pollen from plants that humans use for food or medicine. Other bats eat insects that damage crops.

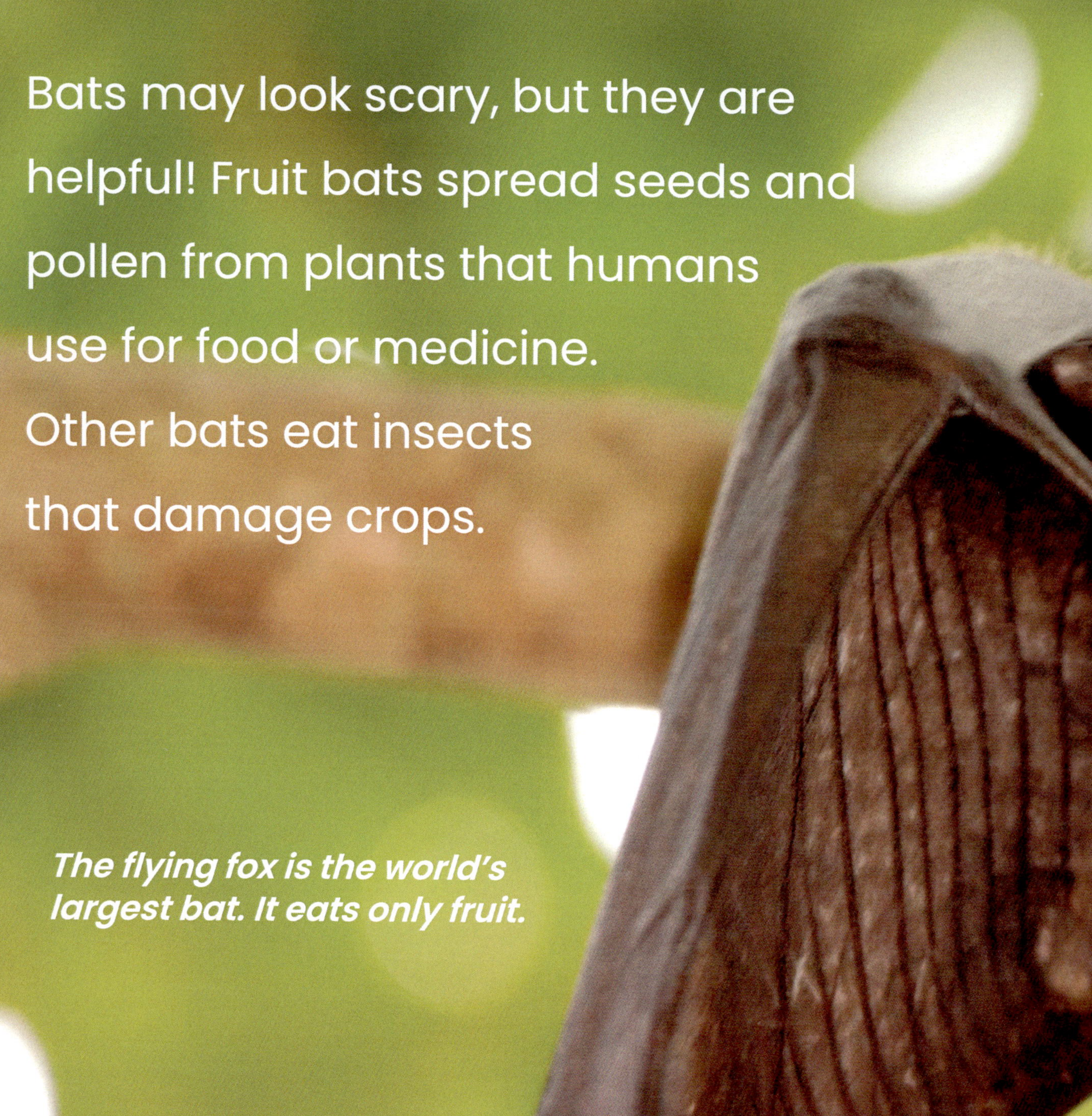

The flying fox is the world's largest bat. It eats only fruit.

Bats have good eyesight, but they hear even better. They use **echolocation**, or sound images, to find food in the dark.

Bats bounce clicks and squeaks off objects. The sounds form sharp images in their brains.

CREEPY OR COOL?

Bats with a noseleaf honk through their noses for echolocation. These odd skin flaps may also help catch echoes.

Mexican fruit bat

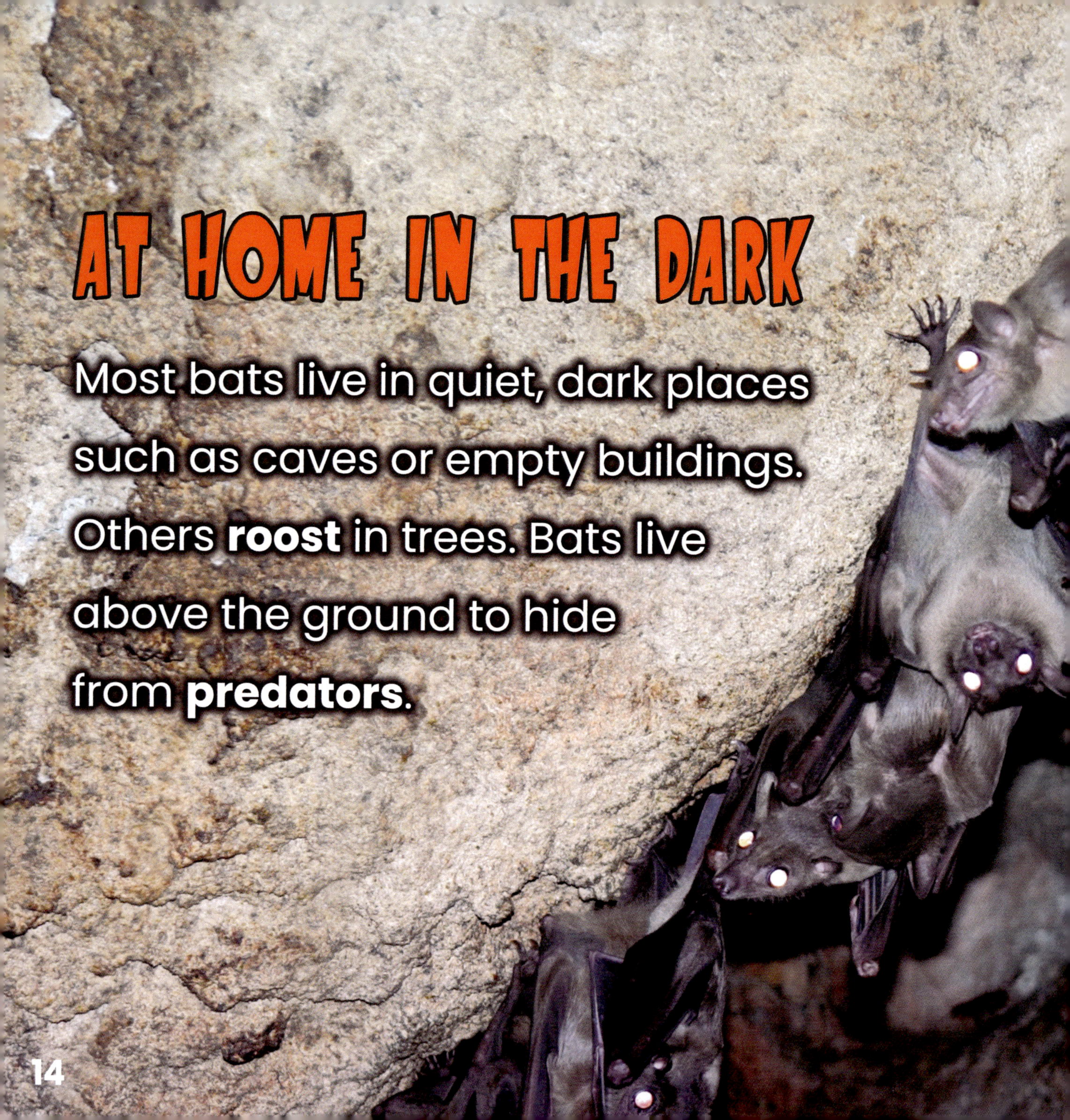

AT HOME IN THE DARK

Most bats live in quiet, dark places such as caves or empty buildings. Others **roost** in trees. Bats live above the ground to hide from **predators**.

A group of bats living together is called a colony.

Most bats sleep upside down. They hang onto branches and ledges with hooked toes.

Some bats wrap their wings around themselves to keep warm.

Bats live nearly everywhere except very cold places. Some bats **migrate** to warmer places in the fall.

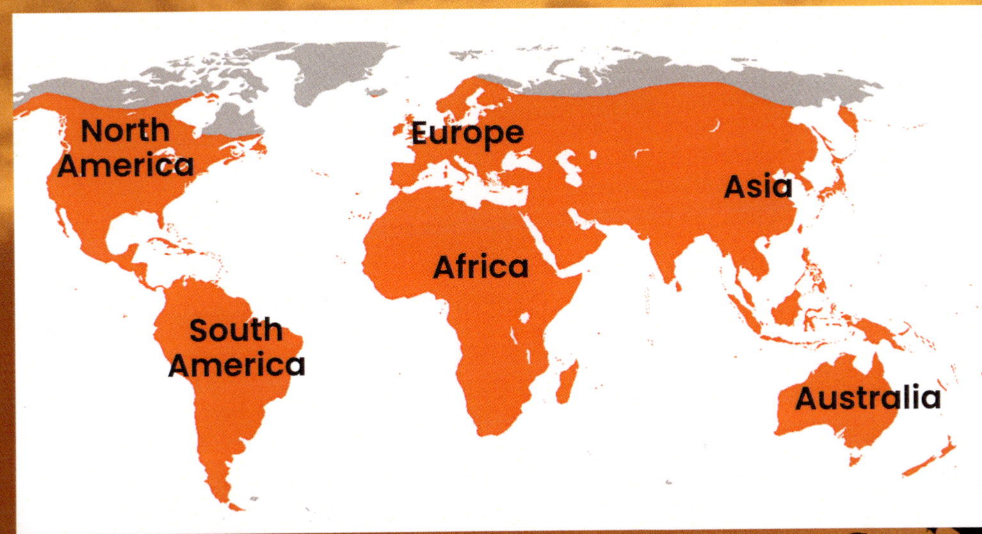

Areas of the world where bats can be found

BABY BATS

Most mother bats give birth to one pup each year. Helpless bat pups cannot fly at birth. They cling to their mothers.

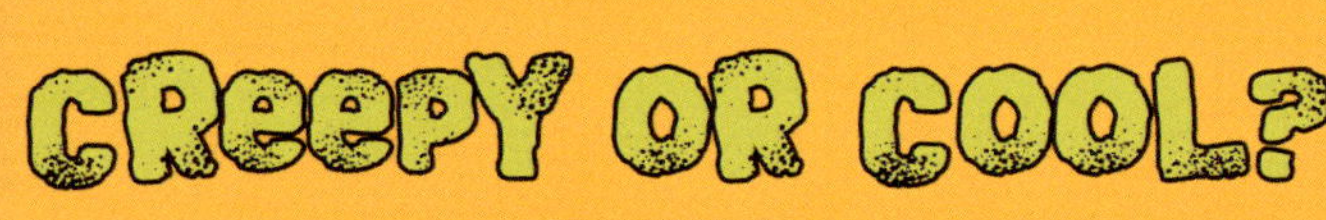

Almost all bats are born without hair, and with their eyes closed.

This gambian epauletted fruit bat is flying with a baby on her belly.

Pups quickly grow too heavy to fly with their mothers. After four to six weeks, bat pups can fly and hunt on their own.

ARE BATS ENDANGERED?

Bats are important to the planet. People have destroyed bat **habitats** or killed bats from fear.

Since 2006, white-nose fungus has killed more than 1 million bats in North America. Scientists are trying to find a way to stop it.

Scientists study and collect information on bats to help bats survive.

NAME THAT BAT!

Match each bat name below with the correct photo:

- -Flying fox
- -Mexican fruit bat
- -Gambian epauletted fruit bat
- -Greater bulldog bat
- -Wrinkle-faced bat

Answers: **a.** Gambian epauletted fruit bat, **b.** Wrinkle-faced bat, **c.** Greater bulldog bat, **d.** Flying fox, **e.** Mexican fruit bat

GLOSSARY

echolocation (EK-oh-loh-KAY-shun): Sending soundwaves and using their echoes to "see" where things are

habitat (HAB-i-tat): A place where an animal usually lives

mammals (MAM-uhlz): A group of animals that are warm-blooded, have backbones, and feed their young milk. Most also have hair or fur.

membrane (MEM-brane): A thin skin or covering layer

migrate (MYE-grate): To move from one region to another with the seasons

nocturnal (nahk-TUR-nuhl): Active at night

predators (PRED-uh-turz): Animals that hunt other animals for food

roost (ROOST): A place where winged animals rest

tropics (TRAH-piks): Warm or hot areas of Earth

INDEX

School-to-Home Support for Caregivers and Teachers

This book helps children grow by letting them practice reading. Here are a few guiding questions to help the reader build his or her comprehension skills. Possible answers appear here in red.

Before Reading

- **What do I think this book is about?** *I think this book is about flying bats. I think this book is about where bats live.*
- **What do I want to learn about this topic?** *I want to learn if bats attack people. I want to learn more about bats and caves.*

During Reading

- **I wonder why...** *I wonder why bats sleep upside down. I wonder why bats wrap their wings around themselves when they sleep.*
- **What have I learned so far?** *I have learned that bats use echolocation, or sound images, to find food in the dark. I have learned that bats help people by spreading seeds and pollen from plants that humans use for food or medicine.*

After Reading

- **What details did I learn about this topic?** *I have learned that most bats are nocturnal, they fly at night and sleep during the day. I have learned that most bats eat mosquitoes and insects.*
- **Read the book again and look for the glossary words.** *I see the word* ***membrane*** *on page 5, and the word* ***nocturnal*** *on page 8. The other glossary words are found on page 23.*

Crabtree Publishing

crabtreebooks.com 800-387-7650

Print book version produced jointly with Blue Door Education in 2022

Content produced and published by Blue Door Publishing LLC dba Blue Door Education, Melbourne Beach FL USA.

Written by Tracy Nelson Maurer
Production manager: Candice Campbell

Hardcover 978-1-4271-6160-4
Paperback 978-1-4271-6172-7

Printed in the U.S.A./CP052026

Library and Archives Canada Cataloguing in Publication
Title: Bats / Tracy Nelson Maurer.
Names: Maurer, Tracy Nelson, 1965- author.
Description: Series statement: Creepy but cool | "A Crabtree seedlings book". | Includes index. | Previously published in electronic format by Blue Door Publishing FL in 2015.
Identifiers: Canadiana (print) 20210200731 | Canadiana (ebook) 2021020074X | ISBN 9781427161604 (hardcover) | ISBN 9781427161727 (softcover) | ISBN 9781427161840 (HTML) | ISBN 9781427161963 (EPUB) | ISBN 9781427162083 (read-along ebook)
Subjects: LCSH: Bats—Juvenile literature.
Classification: LCC QL737.C5 M39 2022 | DDC j599.4—dc23

Published in Canada
Crabtree Publishing
616 Welland Avenue
St. Catharines, Ontario
L2M 5V6

Published in the United States
Crabtree Publishing
347 Fifth Avenue
Suite 1402-145
New York, NY 10016

Photo credits: www.Shutterstock.com and www.istock.com. Cover © Ricardo Reitmeyer, page 4-5 © Matteo Volpi, bat skeleton © Waddell Images: page 6-7 © Ivan Kuzmin. page 8 © Ivan Kuzmin. page 9 top photoc © Dr. Morley Read, bottom photo © Jplevraud. page 10-11 © Stephane Bidouze, page 12-13 © Jouke van Keulen, inset photo © Ivan Kuzmin. page 14-15 © George Burba, page 16 © jakit17, page 17 © Sarun T, page 18-19 © Ivan Kuzmin, baby bat photo © Gert Vrey. page 20 photo © U.S. Fish and Wildlife Service page 21 © Chokniti Khongchum

Library of Congress Cataloging-in-Publication Data
Names: Maurer, Tracy Nelson, 1965- author.
Title: Bats / Tracy Nelson Maurer.
Description: New York : Crabtree Publishing, [2022] | Series: Creepy but cool - a Crabtree seedlings book | Includes index.
Identifiers: LCCN 2021018427 (print) | LCCN 2021018428 (ebook) | ISBN 9781427161604 (hardcover) | ISBN 9781427161727 (paperback) | ISBN 9781427161840 (ebook) | ISBN 9781427161963 (epub) | ISBN 9781427162083
Subjects: LCSH: Bats--Juvenile literature.
Classification: LCC QL737.C5 M387 2022 (print) | LCC QL737.C5 (ebook) | DDC 599.4--dc23
LC record available at https://lccn.loc.gov/2021018427
LC ebook record available at https://lccn.loc.gov/2021018428